Contents

D1344108

Introduction

The Statutory Framework for the Early Years Foundation Stage (2012) requires providers to ensure there is 'access to an outdoor play area or, if that is not possible, ensure that outdoor activities are planned and taken on a daily basis'. In other words, there is an assumption that children will spend at least some of every day outdoors, and that this will apply in rain, wind and snow as well as in sunny summer weather. Of course, there will be times when it's too wet, windy or snowy for children to be out, but these occasions are likely to be exceptional.

This collection of 50 fantastic ideas for rain, wind and snow has been put together with the above guidance in mind. The aim is to provide practitioners with a bank of ideas that they can use and adapt to their own circumstances. Some of the suggested activities are old favourites with which some practitioners may already be familiar. Others are less well known and some are, as far as we know, original. Even the tried and tested favourites have additional suggestions in the 'Taking it forward' section, which will give them a new slant. The emphasis throughout is on problem solving and enquiry. Most of these activities can be completed by children on their own, once they've been helped to understand what's required. Other aspects will require adult supervision, and some are for adults only, with children as observers.

The structure

Activities on pages 8 – 25 are for wet weather, 26 – 37 for windy weather, 38 – 54 for frost, ice and snow and 55 – 64 for any weather. However, there is an overlap; some of the activities in every section can be tackled in different weathers, so don't feel constrained by our suggestions.

All the activities follow the same structure

'What you need' lists the resources required for the activity, and usually for the suggested extensions too. These are basic resources, and are likely to be found already in most settings. Where more unusual items appear we've tried to suggest sources. We have also suggested websites for downloading templates, etc. We recommend that you check the 'What you need' list well before embarking on any activity with the children, as well as reading through what the items are needed for.

'Top tip' boxes offer a brief suggestion, warning or piece of advice for tackling the activity. Usually these are things we wish we had known before we did them ourselves!

Next comes a 'Health & Safety' warning. In many cases these are obvious, and sometimes we say there are no specific hazards involved in completing that activity. However, this doesn't mean that normal health and safety procedures should not be followed.

'What to do' gives step-by-step instructions for completing the task(s). Please read all these before you start. Again, there are sometimes internet references, and in one case a link is given to a YouTube video which demonstrates the instructions we have given. All these sites were active and available at the time of writing, although the internet is constantly changing and we can't guarantee they'll be there in the future.

'Taking it forward' contains some more ideas for additional activities on the same theme. These will be particularly useful for things that have gone especially well. In many cases they use the same resources, and in every case they have been designed to extend the children's learning and broaden their experiences.

Finally, 'What's in it for the children?' is a brief statement that indicates how the activities contribute to learning.

Handling tools and materials

Adult supervision

Reference has already been made to the fact that some of the activities need an adult to take over if they are to be completed safely. For example, anything that involves handling hot items or substances, or cutting out using sharp knives. Practitioners will know which tasks can be safely accomplished by which of their children, and all settings will have health and safety procedures in place that should cover everything suggested in this book.

Outings

For some activities, we also suggest places where children should be taken. Readers will know their own areas and the locations that are suitable, and will also understand the requirement for the permission of parents or carers for children to take part in any outings. In the words of the EYFS Framework (2012), 'Children must be kept safe while on outings, and providers must obtain written parental permission for children to take part in outings. Providers must assess the risks or hazards which may arise for the children, and must identify the steps to be taken to remove, minimise and manage those risks and hazards. The assessment must include consideration of adult to child ratios. The risk assessment does not necessarily need to be in writing; this is for providers to judge.'

Natural materials

Some activities involve working with mud, sand, stones twigs and sticks. Teach children not to put their fingers in their mouths, to wash their hands after using natural materials, and never use these things to hurt or endanger others.

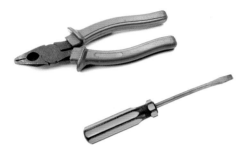

Tools

Some of the activities in this book may need you to expand the range of tools available to the children in your setting. We would advise you to buy the best quality you can afford, as these will last longer and be safer than cheaper versions. Real tools (small hammers, screwdrivers, saws, pliers and safe knives) are fascinating for young children, and can be built into your daily programme as long as you spend time at the beginning explaining the risks to children and training them in their safe use.

Books

The following books have more ideas for things to do outside in wet, windy or rainy weather:

The Little Book of Outdoors in All Weathers

The Little Book of All Through the Year

The Little Book of Growing Things

The Little Book of Maths Outdoors

The Little Book of the Seasons

Books in the 50 Fantastic Things series

All from **www.bloomsbury.com/featherstone**

Food allergy alert

When using food stuffs to enhance your play opportunties, always be mindful of potential food allergies. We have used this symbol on the relevant pages.

FOOD allergy !

Skin allergy alert

Some detergents and soaps can cause skin reactions. Always be mindful of potential skin allergies when letting children mix anything with their hands, and always provide facilities to wash materials off after they have been in contact with skin. Watch out for this symbol on the relevant pages.

SKIN allergy !

Safety issues

Social development can only take place when children can experiment and take reasonable risks in a safe environment. Encouraging independence and the use of natural resources inevitably raises some health and safety issues; these are identified where appropriate.

Children need help and good models for washing their hands when using natural materials or preparing food. They may need reminding not to put things in their mouths, and to be careful with real-life or found resources.

SAFETY FiRST!

Wet weather walk

No need to stay in — get your coats and go walking in the rain.

What you need:

- Boots, raincoats, hats, umbrellas
- A few spares in a bag or backpack
- A camera or smartphone

Top tip ⭐

Involve the children in planning where to go for your walk, but have a route in mind where you'll be able to see a range of wet weather effects.

Taking it forward

- Talk about what happens when it rains too much. Download some pictures of floods.
- Read or tell the story of Noah and the ark. The children could play it out with small world animals.

What's in it for the children?

A wet walk encourages observation, and thinking about how we cope when it rains.

Health & Safety

Don't go when it's too wet – gentle rain is best because you can stay out longer. Make the necessary arrangements to take the children away from the setting.

What to do:

1. Tell the children that you're going for a walk in the rain and discuss with them what it would be sensible to wear.

2. Plan where you're going to go. Involve the children in this, too. Talk about what you might see.

3. When you go out, start by encouraging the children to feel the rain on their faces and hands.

4. Help them to notice the different things that happen when it's raining. Talk about the sights and sounds – tyres splashing on a wet road, water running down drain pipes and gutters and dripping from trees. Where does all the water go?

5. Find some puddles and look for reflections. Enjoy splashing through them. Squelch on wet grass.

6. Take some photos of your walk to make a record and to talk about later.

7. When you get back and have dried off, download your photos onto your interactive whiteboard (IWB), if you have one, and talk about the walk. Get the children to describe what they saw and heard. Get them to imitate some of the sounds. How did they feel when they were in the rain? What did they enjoy and not enjoy?

8. Make a wet–walk display of their drawings and the photos.

Making a wet den

It's great fun being in a tent in the rain, so why not make your own?

What you need:

Some of ...

- **Plastic sheeting, shower curtains, waterproof fabrics**
- **Sticks, poles, canes to support the sheeting**
- **Buckets or large cans filled with sand or stones** (to support the sticks)
- **Large stones, bricks or weights** (to hold the edges of the sheet down)
- **Stout string and clothes pegs, cable ties, duct tape**

Top tip ⭐

Your shelter is meant to be enjoyed in the rain, but it's best to build it when it's not raining!

Taking it forward

- Talk about who uses tents – e.g. campers, circuses – and what they're used for – e.g. shelter, weddings, shows.
- Make some very small tents for toys, and use these for small world play and storytelling.

What's in it for the children?

Shelter is a basic need, and it's not that easy to make your own. This activity gives lots of opportunities for thinking about what's involved and solving problems.

➕ Health & Safety

Sticks and canes have sharp ends and bricks and big stones are heavy. Ensure children are aware of the risks, without stifling their enjoyment.

What to do:

1. Discuss with the children where to build their den.

2. You can either make a frame of your own, using some of the resources listed, or if you have a climbing frame you could fix the sheeting to that. To make your own frame, fill the buckets or cans with sand and stand the sticks or canes in them – or drive them into grass. An 'A' shaped frame works best because the rain will run off.

3. Use the string and pegs to fix sheets together and secure them to the frame.

4. Don't forget that you'll need to save some of the plastic to make a groundsheet.

5. Sit in your den in the rain. Listen to and talk about the sound of the rain on the sheeting. Are there any leaks? Where? How can they be stopped?

6. Add some sleeping bags and camping equipment.

Waterways

Make the water go where you want – it's not always easy!

What you need:

- Buckets
- Lengths of plastic guttering, pipes and tubes
- Duct tape to join these together (NB. some brands won't stick to things that are already wet)
- Bricks and blocks to prop up the waterway
- Bowls and trays that will contain water
- A hose if you have an outdoor tap
- Plastic jugs and funnels

Taking it forward

- Construct a waterway that will collect rainwater from guttering or a downpipe on your building. It might be used to fill a butt for watering the garden.
- Download some pictures of waterwheels. Talk about what they're use for. You might like to make one – see **http://howto. wired.com/wiki/Build_a_Plastic_ Cup_Waterwheel** for instructions.

What's in it for the children?

This activity gives children opportunities to experiment with water and observe the way it behaves.

Health & Safety

When children are absorbed in an activity they may not be aware they're getting cold, so be ready to remind them when to come indoors.

What to do:

1. Talk about what you're going to do – make some waterways. Look at the resources you have collected and discuss how they can be used.

2. Give the children time to build their own waterways using gutters, tubes and pipes, and to observe how the water behaves as it flows through them.

3. Make some simple card or plastic boats (or use ready-made ones) and float them down the waterways. You could have races.

4. Use sticks, twigs, sand, or soggy newspaper to make dams.

5. When the children have played and experimented for a while, set them a challenge to channel water from one place to another.

6. Make a waterfall with guttering. How high can you make it? What can you use to collect the water at the bottom? Let the children feel the force of the falling water at the top and at the bottom of the waterfall. What's the difference?

7. Photograph what goes on to use in discussions and displays later.

Top tip ⭐

This doesn't have to be done in the rain but it's fun if it can be. Whether it's raining or not, children will get wet, so make sure they're dressed with this in mind.

Wonderful wellies

Have fun in the mud with a selection of wellies!

What you need:

- A muddy area: this may be easy to find in your grounds, or you can make your own mud patch in a builders' tray
- Wellingtons – lots of different ones with a variety of tread patterns
- A spade or a flat piece of wood
- Rolls of paper
- Paint (various colours)
- A hose

Taking it forward

- Try to walk with two left boots, or two right boots. How does it feel?
- Talk about animal tracks, and download some examples (try **http://www.naturedetectives. org.uk/download/hunt_tracks. htm**, which has downloadable identification sheets, as well as lots of other ideas for things to do with mud). If you have some open ground or woodland nearby go for a walk and see if you can find any (you'll probably also find dog and bird tracks).

What's in it for the children?

This activity provides plenty of practice in experimenting and observing, as well as the chance to get creative.

 ### Health & Safety

Children will be handling muddy wellies and will get mud on their hands. Discourage them from getting it in their mouths, especially if you have any habitual suckers and chewers.

What to do:

1. Put all the wellies in a pile. The children take two each (not necessarily a pair).

2. Take the children outside to the mud patch and let them take turns to stamp through the mud. Look at the different tracks made by the boots.

3. Use the spade or piece of wood to smooth out the mud and try again. Encourage them to make patterns.

4. Try hopping, jumping and skipping, and look at the differences.

5. Drop blobs of different coloured paints on the mud and see what happens when children walk through them.

6. Make welly print patterns on paper. If there are some wellies that are too small for the children's feet they can put them on their hands. Let them experiment, but be ready to make suggestions if you think they need more ideas (straight and wavy lines, circles, stars, spoked wheels).

7. Watch what happens to the mud and the paint as it rains.

Top tip ★

Ask parents for old wellies. Look in lost property, try car boot and rummage sales.

Passionate about puddles
Children love these fascinating, instant mini-lakes.

What you need:

- Clothing and footwear for wet weather
- Chalk
- Meter measuring stick: you could make one from a broom handle or a cane marked with coloured lines
- Camera or smartphone

Top tip ⭐

You can test predictions of where puddles form by making artificial puddles with a hose.

Taking it forward

- Do puddles always appear in the same places? Talk about why this might be.

- Have a puddle jumping competition, when each child finds the widest puddle they can jump. Be prepared for everyone to get very wet!

What's in it for the children?

This activity gives children practice in testing predictions through observation, in making deductions and in measuring.

✚ Health & Safety

Puddles can hide glass, nails and other objects which might cause harm. Check them out first.

What to do:

1. Go outside when it's dry and predict where there might be puddles when it rains. Take photos of places where the children predict the puddles will be.

2. Involve the children in watching the weather forecast and the local signs for rainy weather. Have everything ready to go outdoors quickly when the rain comes.

3. After it's rained for a while, go outdoors to look for puddles. Are they in the places predicted? Use your photos to check. Are there some puddles in other places? What can you say about the sorts of places where puddles form?

4. Look carefully at each puddle and see if you can tell how deep it is. Use your measuring stick to measure depth and distance across each puddle. Walk through the puddles and count the steps taken to cross each. Stand in the middle and mark the water level on children's boots.

5. When the rain stops, explore evaporation. Draw a chalk line around some puddles and return later to see if any have got smaller. What has happened to the water? Keep drawing lines as the puddle evaporates. If the rain holds off you might be able to do this over several days and record how long it takes for various puddles to disappear. Do they go down faster when the sun shines?

Tunnel to Bookland
Make a den with a difference!

What you need:

- A pop-up tent
- A plastic play tunnel, big enough for children to crawl through
- Duct tape
- Cushions or small beanbags
- A basket of books
- A CD or MP3 player, or a radio
- Some hair scrunchies or wide, coloured elastic bands

What to do:

1. Put the tunnel so the end is just inside the door to your outside area.

2. Put the pop-up tent at the end of the tunnel and put the cushions and book basket inside. Add other things such as a music player or CDs.

3. Link the end of the tunnel to the tent with strips of duct tape or string. Make sure the door of the tent is open all the time – tape or tie it back and tell the children not to close it.

4. Decide how many children can play in Bookland at one time – it will depend on your children and the size of the tent. Put that number of scrunchies or bands on a hook, or in a basket, near the entrance to the tunnel. This makes it easy for adults and children to regulate the numbers.

5. Make sure the space is not dominated by particular children, and talk about the way children should behave when they are in Bookland.

6. This sort of space that is too small for adults is particularly exciting for children. Make time to talk about what it is like to be in a tent in the rain.

Top tip ★

Set this up before the session, so you don't get wet!

Taking it forward

- Add some child-sized sleeping bags to make the tent extra comfortable.

- Everyone will want a turn, so let the children decide how to regulate the time for a turn in Bookland, and how you will all know when it's time to change over.

What's in it for the children?

Being in this space is very exciting, as children will need to regulate their own behaviour and wait for a turn.

Health & Safety

Make sure the tent is within earshot of the adults.

Charming worms

Worms aren't nasty. Explore and understand them better.

What you need:

- Some string and four tent pegs or sticks
- A watering can
- Some mustard powder
- A wooden spoon
- Tweezers or kitchen tongs
- A shallow bowl or tray
- Magnifying glasses, water in spray bottles, paper towels
- Camera or smartphone

What to do:

1. This activity is suitable for a showery day, when the worms are likely to be near the surface.

2. Talk with the children about what you are going to do, and identify any children who may be frightened of worms, or don't understand how to look after them.

3. Make sure the children know that earthworms need to be damp at all times, and that picking them up will remove the invisible protective coating on their skins. Never make a child hold something that scares them, but encourage them to look.

4. Go outside and choose a place for your worm experiment – grass or soil surfaces will be ideal. Measure off a square of grass or soil about a metre on each side, and mark it with string and pegs. This will be your 'worm experiment area'.

5. Put some cold water in a shallow tray.

6. Put four litres of cold water in your watering can and add 40 grams (¼ cup) of dry mustard powder. Help the children to stir it with the wooden spoon until the mustard is dissolved. The mustard is completely safe for children to use, but it tastes horrible!

7. Slowly pour the mustard solution over the soil inside the boundaries of your study area. Pour it so it soaks into the soil instead of running off. The worms will start coming up. Don't worry; the mustard does not harm the worms, it just makes them escape to the surface.

Taking it forward

- When you have had a look at the worms, return them to the ground near the place where they came from, but outside the square where you put the mustard.

- Use the internet to find out more about worms, and make a display.

What's in it for the children?

This activity gives children an opportunity to examine live animals and learn how to handle them.

Health & Safety

Worms are living things. Teach children how to handle them carefully.

Top tip ⭐

Worms are becoming rarer. If you have trouble finding them, see if you can locate a molehill and look there.

8. Now watch what happens. It won't take long for the worms to come up, and you can pick them up very gently with tweezers or tongs. Quickly put them into the fresh tap water to rinse off the mustard solution.

9. Take the worms out of the water, put them on a wet paper towel, and use a magnifying glass to look at them. Explain that the worms are 'breathing' oxygen through their wet skin, so they must be kept moist at all times. Provide some water sprays for children to use, and explain that because worms are fragile animals and can be hurt easily they must be handled with gentleness.

10. Take plenty of photos!

Plastic regatta

These boats are simple and quick to make and will float anywhere.

What you need:

- Plastic bottle (1) **with cap**
- **Small square** (about 15 x 15 cm) of thin foam, or a sheet (A4) **of stiff paper or thin card**
- **Rice, sand or gravel** (one cup)
- **Pencil**
- **Sharp pointed scissors and a hole punch**
- **Sheet of thin paper**
- **Duct tape**
- **Blu-Tak**

Top tip ⭐

Make the boats indoors, then when it rains take them out and race them across puddles.

Taking it forward

- Sometimes the wind will blow the boats across the puddles sideways. Can you do anything to stop this?
- Do an internet search for 'plastic bottle boats' and see large boats made of bottles that will carry people.

What's in it for the children?

This is a simple craft activity that works well, gives a sense of satisfaction, and gets children to think about the principles behind making a successful boat.

Health & Safety

Some children may need help with the cutting, particularly because plastic bottles are flimsy and the scissors can slip.

What to do:

1. Wash out the plastic bottle, clean off the label, and leave it to dry.

2. When the inside is completely dry, roll a thin sheet of paper into a funnel and pour the cup of rice into the bottle. It should be about one third full. Replace the lid, tighten it and seal with tape.

3. With the scissors make a small hole (no bigger than the diameter of the pencil) about half way along the bottle.

4. Gently push the pencil through the hole so it stands up to make a mast. Seal the hole with Blu-Tack if it's wobbly. Shake out the rice so it's evenly spread along the bottom of the bottle.

5. Bend (don't fold) the foam or card to line up two edges. Punch a hole about 1.5 cm in from these edges and about half way along. Thread the two holes onto the pencil and open up the foam or card to make a sail.

6. Make several of these boats. You can colour or decorate the sails, and add flags made from coloured tape to the mastheads. Make some with two masts and sails.

7. Wait for the rain and the puddles, and hold your regatta.

Keeping dry

Why do some clothes keep the rain out and others let it through?

What you need:

- Wellies and plastic rain coats
- Some non-waterproof clothing, e.g. T-shirts, shorts, sweatshirts
- Small watering can of tepid (not cold) **water**
- Large plastic bowl or tray
- Large plastic mat
- Towels, mop and bucket

Top tip

This is a messy activity! Do it all outside if possible, or work on a large mat or a bath towel.

Taking it forward

- Sing some rainy day songs.
- Talk about who needs waterproof clothing (firefighters, sailors, police, builders, farmers).

What's in it for the children?

This activity encourages children to think about the nature of materials and testing them, giving practice in using language to describe the action of water.

✚ Health & Safety

Watch out for children slipping on wet surfaces.

What to do:

1. Introduce all the clothing and ask the children to sort the items into those which keep out water, and those which don't.

2. Ask the children how you could find out whether they are right, and how you could find out which clothes keep the water out best. You might hint at the watering can.

3. Put the big tray and the water in the middle of a clear space (outside would be best) and invite a volunteer child to put on the plastic raincoat and the wellies and stand in the plastic tray. Get another child to help you gently trickle water from the watering can over the arm of the coat.

4. Ask the child to take the coat off. Is their arm dry? Let some of the others have a go, but they must be careful pouring the water.

5. Now try pouring water onto a non-waterproof item, such as a T-shirt. Compare the T-shirt with the raincoat. If the children don't use the word naturally, introduce the word 'waterproof' and talk about why some items of clothing are waterproof.

6. Ask different children to dry the coat and boots with a towel. Are they easy to dry? What about the T-shirt?

7. Get the children dressed in their waterproofs and go out in gentle rain. Look and talk about what happens when the raindrops land on their waterproofs. Encourage words such as 'trickle', 'drip', 'drop' and 'bead'.

Chasing rainbows

Rainbows are fascinating, partly because they're soon gone!

What you need:

- **Cooking oil, decanted into clean washing-up liquid bottles** (have several, depending on the number of children)
- **Several plastic trays** (trays from storage units)
- **Plastic dishes** (the ones supermarkets use for soft fruit)
- **Food colouring – red, green, blue and yellow**
- **Camera or smartphone**

What to do:

1. Rainbows appear when there's sun and rain together. Keep on the lookout, and be ready to go out as soon as one appears. Ask the children to name the colours. Sing the rainbow song: you can find the words at http://kids.niehs.nih.gov/games/songs/childrens/singarainbowmid.htm

2. Can you spot a double rainbow?

3. Find some more rainbows. Go to a car park when it's raining and see if you can find any on the ground (where cars have dropped oil).

4. Make your own rainbows. Leave some plastic trays out in the rain so they get full of water. (If it's windy you might have to weigh them down with stones or bricks.) Tell the children they are going to make rainbows in the water.

5. Drip four or five drops of oil from the washing-up liquid bottles into the water and watch what happens. Not too much oil, so children may need to be told to restrain their squeezing!

6. Get them to talk about the colours and effects they see. Gently disturb the surface of the water with a stick, swirling the oil.

7. Next, fill your clear plastic container with clean water – collected rainwater or from the tap.

8. Pour a little cooking oil into a dish. Add some food colouring – enough to make a good, strong colour – and mix well. Make several of these, using different colours.

9. One by one, tip the coloured oils into the water and watch the result. Allow time to see the effect of each one before adding the next.

10. Take plenty of photos, so the children can revisit and talk about the activity afterwards, and to make a display. It will be even better if you can video what happens.

Top tip ⭐

Use the cheapest cooking oil you can find - it works just as well as the more expensive stuff.

50 fantastic ideas for rain, wind and snow

Taking it forward

- Try mixing some more oils with water – olive oil, baby oil, sun lotion. Do they ever mix? Get children to suggest words for how the oils feel on their hands and fingers. Does water feel the same?

- Quarter fill a small, clear bottle with water and add food colouring, enough to make a strong colour. Add the same amount of oil. Put the lid on and shake it up. Watch what happens as it settles and the oil and water separate. Let the children repeat this, observe and talk about it. Add a few drops of washing-up liquid to the bottle and see what happens when you shake it up again.

- Take another bottle and this time add about twice as much oil as coloured water. When the oil and coloured water have separated, drop in half an Alka-Seltzer tablet and watch a dramatic eruption.

- Make a rainbow column. There are instructions at **http://www. stevespanglerscience.com/lab/ experiments/seven-layer-density- column** The children will need adult help and careful supervision, but it's fun to do. Bear in mind these are American instructions – 'Karo syrup' is corn syrup and 'Dawn dish soap' is blue washing-up liquid.

What's in it for the children?

This activity gives opportunities for observing and talking about the properties of liquids.

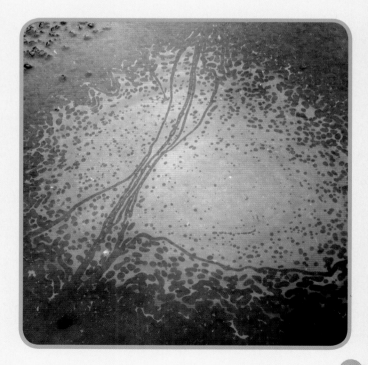

➕ **Health & Safety**

Wet surfaces can be slippery, warn the children, and supervise carefully.

Soak it up

Look at what happens when different fabrics get wet.

What you need:

- A good range of fabrics - cotton, fleece, lycra, leather, net and gauze, rubber, velvet, suede, polyester, towelling, etc. You'll need several of each, and make them about the same size – say 25 cm square
- Plastic buckets
- Plastic cups
- Plastic trays (polystyrene trays from supermarket products)
- A clothes line and pegs

Top tip ⭐

This activity makes a good follow-on from 'Keeping dry' on p17. Most of this activity can be done outdoors in the rain, but you'll need to come in for the last bit. If the weather's too bad you can do the whole thing indoors using bottles to pour the water, but it's less fun.

Taking it forward

- Talk about really absorbent materials such as cotton wool, towels, kitchen roll and babies' nappies. You could test and rank some of these too.
- Watch drops of water or rain as they run down windows and doors, and collect words to describe them.

What's in it for the children?

Predicting the outcome of an experiment is an essential part of scientific method – so is classifying and sorting.

What to do:

1. Before you go out, let the children handle and examine the fabrics. Feel them, scrunch them up and talk about them. Make sure everyone handles all of them.

2. For each of the fabrics ask the children to tell you whether it will soak up water or won't soak up water. Put them in two piles according to what they decide, with a third pile for those they're not sure about.

3. Put some samples from each pile in the bottom of small trays and take them out into the rain, so the children can see what happens as they get wet. Were the children's predictions right?

4. Ask open questions such as 'Where has the water gone?', 'Why do you think the water is staying in drops on this fabric?', 'Why is it soaking into that one?'and 'Why do we use soaking-up or absorbent fabrics for some things?' What about the fabrics they were not sure about?

5. Put the fabrics in a bucket of water (or leave them on a plastic tray in the rain while you come in and get warm!). Let them get good and soggy.

6. Take them out of the water and peg them on a line till most of the dripping has stopped (about five minutes).

7. One by one squeeze each of the fabrics into a plastic cup. Keep each sample with its own cup and use the amount of water collected to rank them in order of absorbency.

8. Remember to take photos of these experiments.

Exploding paint

Make some of these paintballs and have some fun in the rain.

What you need:

- Powder paint
- Flour
- Kitchen roll
- Plastic spoons
- Elastic bands
- A place to throw the paintballs, where you won't upset anyone
- A camera
- A hose for cleaning up

Top tip ⭐

Add a bit of rice to the paintballs - they'll go much better with the extra weight.

Taking it forward

- Line a paddling pool with paper and pour in some water to make the paper damp — not too wet. Make some more paintballs and drop or throw them onto the paper.
- Wait for the paper to dry and carefully take your picture out of the paddling pool. How does it look?

What's in it for the children?

This activity is pure fun in the rain, and will allow children to let off steam in bad weather when they have been indoors.

 ### Health & Safety

There have to be rules about throwing anything, even in a game. Make sure the children understand what they are.

What to do:

1. Spread a piece of kitchen roll on a table.

2. Put a large spoonful of powder paint and two spoonfuls of flour in the middle of the paper.

3. Show the children how to lift the four corners of the paper, and twist the paper, holding the middle bit in your hand (see photo).

4. Put an elastic band around each 'paintball'.

5. When it's raining, take your 'paintballs' outside. Draw a throwing line with chalk, and throw the balls along the patio or path of your setting or house, or onto walls.

6. When you have used all your paintballs, walk over and talk about which ones worked best. Take some photos.

7. Use the hose to wash away the paint, and sweep up the bits of towel.

8. Or you could go inside and make some more!

9. Experiment with several colours in one ball, with more paint, more flour, or by adding some rice or lentils. Dry dipping the ball briefly in a puddle or bucket of water before you throw it.

Magic in the rain

Let raindrops paint colourful pictures for you.

What you need:

- Card or stiff paper
- Food colouring, in as many colours as you can find
- Small paintbrushes
- Outdoor clothing

Taking it forward

- Dry the sheets and use them for wrapping paper.

- The children can work on small cards or card cut from cereal packets, instead of big sheets. They can then have one each, and when they are dry turn them into greetings cards.

- Try wetting the card first and dropping or spraying the colours onto it.

What's in it for the children?

Children will be fascinated by the changes the rain makes to their dots, and by how colours combine to make others.

Top tip ⭐

You can do this on a sunny day and use a hose with a sprinkler or a watering can instead of rain. This will give you more control over the finished result.

What to do:

1. Explain to the children what you are going to do: they are going to put some coloured dots on card and watch the rain turn them into pictures.

2. Before you go out, spread out the sheets of paper or card and give each child or group several different pots of food colouring and brushes (smallish ones are best).

3. Show them how to put spots of colour on the card. Stress that they should be spots or dots – they are not trying to make blocks of colour – and they shouldn't be too close together.

4. Get into your outdoor gear and take the cards out into the rain. Put them down on a flat surface (if necessary, use stones to hold them down) and watch what happens as they get wet. It's a good idea not to let them get too soggy!

5. Let the children decide when the artwork is finished. If the cards get too wet the colours will merge into a dull slush, so you'll need to encourage them to say when they're done.

6. Bring the cards inside, dry them flat and make a display.

7. It's all right to let the colours dry a bit before you go out – the drier they are, the longer it will take for the rain to affect them.

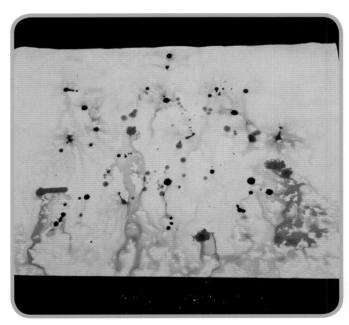

Mud modelling

There's nothing like real mud for messy modelling.

What you need:

- Children's spades and buckets
- Mud
- A builders' tray
- Plastic pots and cups.
- Tin lids, moulds, pastry cutters

Top tip ⭐

When sourcing your mud, be careful to avoid areas where there might be animal faeces. Mud made from a clay soil works best. Get your mud from a clean area and bring it back to your setting to use on a builders' tray.

Taking it forward

- Make a mud fort. The pictures at http://creative.sulekha.com/happy-diwali-making-mud-forts-at-palace-museum-in-baroda_442872_blog will give you some ideas.
- Make some other mud creatures, and when they're dry put them in small world environment. Mud dinosaurs look good in a small world Jurassic Park.

What's in it for the children?

This is a tactile activity which gives children practice is manipulating with their hands.

Health & Safety

It's essential for children to wash their hands after this activity, and to keep their fingers out of their mouths during it.

What to do:

1. Decide in advance where you're going to go for mud. Take the children out, collect the mud in buckets and bring it back to the builders' tray.

2. Let the children play with the mud, squeezing it through their fingers and experiencing the texture. Make sure the mud is nice and squidgy – you may need to add more water if it's not raining enough. Get them to suggest words that describe the mud and how it feels.

3. Encourage the children to pat the mud flat or roll it out, and cut out shapes with the pastry cutters. Try pressing shapes or patterns into the tops.

4. Fill some of the moulds with mud and turn them out. The mud will come out more easily if you first brush the inside of the mould with a little vegetable oil.

5. Make Marvin the Mudbug:
   - Make a fist-sized ball of mud. This is Marvin's body.
   - Make a smaller ball for the head.
   - Find six short sticks and fit them as legs, three on each side of the body. Attach the legs, three on each side of the body.
   - Use a two pebbles for eyes and dry grass for feelers.

6. Bring all your mouldings and mudbugs inside to dry (dry them slowly or they'll crack). When they're dry, paint them and make a display.

Washing day

Take some of your equipment outside for a wash.

What you need:

- Buckets
- **A hose** (if you have an outdoor tap)
- Sponges, mops, washing-up brushes, cloths
- Lots of warm water
- Non-allergenic washing-up liquid
- Waterproof clothing
- Towels
- An empty water or sand tray

Top tip ⭐

Make sure the children are well-covered: this is a wet activity.

What to do:

1. Explain to the children that it is washing day, and that anyone who wants to help may get very wet. You will have plenty of volunteers, particularly if the day is warm.

2. Decide what resources you are going to wash. Talk about the sorts of things that you shouldn't put in water, and those that are waterproof.

3. Collect some of the resources and put them near the door.

4. Let the children help to get the buckets and other things you need for the washing day, and put them in an agreed place outside. A path, patio or other hard surface would be good.

5. Good things to include would be Lego and other plastic construction toys, plastic number equipment, dolls, crockery and pans from the home corner, outdoor wheeled toys – bikes, scooters etc. If the children want to go on, they could wash chairs, plastic storage boxes and anything else that is waterproof.

6. Take turns to scrub the objects with cloths and brushes and plenty of bubbly water. Keep topping up with buckets of hot water from indoors.

7. When each item is clean, prop it upside down to drain, then rinse it with clean water from the hose.

8. Leave the clean things outside to dry, and if the rain doesn't stop, get the children to help to bring the equipment inside and put everything on towels or plastic sheeting.

9. Small items like Lego bricks will dry well on a towel, where you can shake the towel every so often to turn them over.

Taking it forward

- Keeping the equipment clean and tidy is a job that children can really take some responsibility for. Individuals and pairs of children can be in charge of tidiness and cleaning in different parts of the room.

- Sometimes when it rains, you could just wheel the toys out of the shed and let the rain do the job for you.

What's in it for the children?

Children love doing this real-life activity, and it gives them a chance to contribute to keeping the setting clean and tidy.

⊕ Health & Safety

Wet surfaces can be slippery. Warn the children, and supervise carefully.

Take a windy walk

Walking in the wind can be exhilarating and exciting!

What you need:

- Clothing for windy weather
- Streamers
- A camera or smartphone

Top tip ⭐

Don't go when it's too windy – a strong breeze is best. Don't stay out too long.

Taking it forward

- Collect windy words.
- In an open space, move your arms like the wind. Do a wind dance, spinning, swirling, whirling and twirling. Make some wind noises to go with the dance.
- Read or tell some stories about the wind: e.g. *How Does the Wind Walk?* by Nancy White Carlstrom; *The Windy Day* by Anna Milbourne; *Feel the Wind* by Arthur Dorros; *Winnie the Pooh and the Blustery Day* by A A Milne.
- Talk about severe winds and search the internet for pictures showing the effects of gales and hurricanes.

What's in it for the children?

A windy walk encourages observation. Children can get thinking and talking about their own experience of the wind and how it affects us.

 ## Health & Safety

Don't ignore wind-chill, which affects children more quickly than adults – proper clothing is essential.

What to do:

1. Talk with the children about the wind. Can you see the wind? How do you know it's there? Can you hear the wind?

2. Tell them that you're going for a walk in the wind and ask them what they might see and feel (smoke from chimneys, flags blowing, tree branches swaying, leaves and litter whirling about, hats, hair and clothing blowing).

3. When you go out, start by encouraging the children to feel the wind on their bodies, faces and hands. Get them to point in the direction the wind is blowing. Let them take a streamer each and see if they were right.

4. Walk and run with the wind and into it. Talk about how different each feels. Which is easier?

5. Remind the children of your discussion about what they might see. Can they see any of the things they predicted?

6. Take photos to show the wind at work and to make a record to talk about later.

7. When you get back and have warmed up, load your photos onto your IWB, if you have one, and talk about your walk. Get the children to describe what they saw, heard and felt. Why can the wind move some things and not others? How do we use the wind?

8. Make a windy walk display of their drawings and the photos.

Time for chimes

Most children love things that make a noise!

What you need:

- Objects that make a noise when banged together – jar lids, shells, metal cutlery (cheaper the better), **small tin trays, buckles, etc.**

- **Two metal coat hangers for each chime**

- **Tape and string**

- **Glue**

- **A piece of card, or a cork or a large feather**

Top tip ⭐

Jam jar lids make a good sound, and so do tins.

Taking it forward

- Hang old saucepan lids and metal kitchen tools in bushes and trees in your setting, or in gaps in fences. Use wooden spoons to 'play' them.

- Make a chime with lengths of bamboo or metal tubes hung on strings of various lengths.

- In a strong wind, put out tubes and glass bottles and listen to the sound as the wind blows across their tops. Experiment with putting different amounts of water in the bottles.

What's in it for the children?

This activity gives children practice in following instructions, and requires manual dexterity.

✚ Health & Safety

Wire hangers and metal objects can have sharp edges. Warn the children of the dangers and supervise them closely.

What to do:

1. Fix the coat hangers together with tape, at right angles to each other. The children will need help with this. If you can cut the hook off one of the hangers it will be easier to fix them together.

2. Hang the metal objects from the hangers with string. Check that they're near enough to strike each other but far enough away for the strings not to tangle. When you're happy, glue or tape them in place.

3. Place one longer string in the middle, with a feather, card or cork to catch the wind.

4. Hang your wind chime in a tree or bush, or from the climbing frame or an overhang. Experiment with different places to get the best result.

Look what the wind's blown in!

The wind plays a big part in spreading seeds and plants.

What you need:

- Clothing for windy weather
- Plastic or paper bags for collecting
- A park or woodland to walk in
- Camera or smartphone
- Fabric or backing paper, stapler and glue (for the display)

What to do:

1. Tell the children you are going on a walk to look for natural objects blown by the wind: leaves, twigs, pine cones, acorns, conkers, seeds, nuts, berries.

2. Get dressed up and give each child a collecting bag. Go out and look for the things you have talked about. Take photos of the children while they quest.

3. When you get back, tip out the bags and compare what you have found. How many different sorts of leaves did you find? Get the children to arrange them in order of size. Do the same with the nuts, berries and seeds. Did you find anything you hadn't talked about beforehand?

4. Spread out the fabric or paper on a flat surface and make a display using the things you found. Print some of the photos you took of the children (or they took of each other) and add them to the display.

Taking it forward

- Look at the differences between leaves. Use books or the internet to find out what trees they came from.

- Most found seeds and berries will germinate. Put some in a plastic bag and put them in the fridge for a few days (to simulate winter). When you take them out, squash the berries onto a sheet of kitchen towel and clean away the skin and flesh to expose the bare seeds. Plant them in small pots, water them and put them outside in a sheltered spot where they'll catch some rain. Go out and look at them every so often to see if there are any shoots.

- Put on some swirly music and imitate the seeds blowing from the trees, landing on the ground, sleeping and growing.

What's in it for the children?

This activity helps children realise the importance of wind in the natural cycle of seeding and growth.

Health & Safety

Some seeds and berries are poisonous. Keep everything away from mouths, and be aware of what children are picking up.

Top tip ⭐

Late autumn is a good time for this activity, but you can do it anytime things are seeding. You can combine it with the windy walk, although it's best treated separately.

Wind wheels

Harness the power of the wind to make some garden decorations.

What you need:

- Paper or thin card
- Scissors
- Glue
- Pencils
- Mapping pins
- Green garden sticks or thin dowel
- Beads (optional)
- Crayons or markers

Top tip ⭐

Best to use crayons or markers for the colour. Paint makes paper wrinkle and the wind wheels won't work as well. Enlarge the whole slightly to make sure the wheel spins.

Taking it forward

- Pin a row of wind wheels on the fence and watch them spin together.
- Make wind wheels from paper plates and foil dishes. Make cuts all round the edges and fold them in. Pin the wheels to fences, posts and trees and watch them spin.
- Fix wind wheels to the handlebars of bikes and scooters.
- Talk about what wind turbines do, and look them up on the internet.

What's in it for the children?

This activity gives children chance to think about how the wind can be harnessed.

What to do:

1. Cut two 15 cm squares from paper or thin card. You can download and print templates for these from **http://www.firstpalette.com/tool_box/printables/basicpinwheel.html**. Glue them together.

2. When they're dry, colour both sides with markers or crayons. With a pencil, draw two light lines diagonally between the corners (these are already marked on the downloadable templates).

3. With scissors, cut along each line. Start at the corners and stop 2.5 cm from where they cross.

4. Bring every other corner to the centre – be sure not to crease the paper and pin to the stick. Leave the pin loose enough for the wind wheel to turn – putting a small bead on the pin will make it easier (children will need help with this bit).

5. Go outside and test it.

Let's go fly a kite

You can make simple kites and windsocks from many things.

What you need:

Paper plate kites

- Large white paper plates
- **Scissors** (a range of sorts)
- String
- Felt pens, ribbon, wool, crêpe paper
- Hole punch, stapler
- Short sticks or garden canes

Windsocks and birds

- Felt pens, ribbon, wool, feathers
- Scissors
- White glue, felt pens
- Paper and coloured, sticky paper
- String
- Empty paper towel rolls

What to do:

Paper plate kites

1. Collect the materials and talk with the children about what you are going to do.

2. Draw shapes, patterns or faces on the paper plate. Punch four holes equally spaced around the edges.

3. Measure and cut four equal pieces of string and tie one through each of the holes. Punch four more holes, between the others. Cut some streamers from ribbon, crêpe paper or wool, and tie them through the holes.

4. Tie the ends of the strings together at the front of the plate (so you can see the pattern as the kite flies). Take another piece of string (not too long but long enough to fly) and tie one end to the knot of four strings and the other to a stick or cane.

5. Now go outside and fly your kites!

Top tip ⭐

Holes can tear in the wind when the strings pull out. If you tape both sides of the plate rim before punching the holes it will make them tougher.

For both these activities it's a good idea to make a sample as the children watch, then leave them to make their own, asking for help if they need it.

Windsocks and birds

1. Collect the materials together and talk with the children about what you are going to do.

2. Paint the paper towel tubes in bright colours.

3. Staple several long streamers onto the bottom of each tube, make a hole in the top and tie a string through it. If you like you can add stickers or glitter to the streamers.

4. Run around outside with the windsocks. Hang them up outside and watch them dance in the wind.

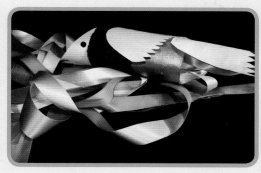

5. After the children have enjoyed the windsocks for a bit, turn them into birds by sticking on beaks cut from card, and wings and tails from cut card or real feathers.

Health & Safety
Whenever things are flying about there's a risk of damage to eyes. Be alert to any hazards.

Taking it forward

- Make kites from other shapes of card – squares, triangles, diamonds. Do some work better than others?

- Make plastic bag kites from carrier bags, by cutting the bottom off the bag and tying the handles together. Cut streamers from strips of other bags, and stick them on with sticky tape.

- Have a competition to see who can keep their kite in the air the longest.

- Sing 'Windy Day' or flying songs as you work and as you fly your kites outside.

- Enter an internet search for 'make your own kite' for lots more kite ideas.

What's in it for the children?
These activities need concentration, plus manual skills for the cutting, stapling and sticking. Running with kites is great exercise.

Dry this

Make use of the wind to dry the washing!

What you need:

- Small items to wash – dusters, face flannels, doll's clothes, socks
- Plastic bowls and soap flakes
- A large jug of warm water
- A clothes line and pegs

Top tip ⭐

The washing is best done outside on a fine day.

Taking it forward

- Take photos at each stage of the washing process. Ask children to help you put them in order and think of a caption for each picture. Make them into a book or frieze to display.

- Ask the children how their parents or carers dry their washing. Construct a chart showing the popularity of different ways. Ask what other methods could be used to dry washing (encourage off-the-wall suggestions). Most of them will involve passing warm air over the things to be dried.

- If you have a washing machine in your setting, have a look at how it works and compare the programmes with your own washing sequences.

What's in it for the children?

This gives children chance to explore another use of the wind.

⊕ Health & Safety

Use mild soap flakes, not detergent. Check for allergies before allowing children to wash clothes, and make sure they rinse and dry their hands afterwards.

What to do:

1. Show the children a dirty cloth as a starting point for discussing why we wash our clothes. Introduce the washing-up bowls, jug of water and other items and let the children explore the washing things. This is a good outdoor activity!

2. Talk about the steps in the process of washing, rinsing and pegging the items to the washing line. Establish a sequence by using 'first', 'next', etc. As you talk go through the process together:
   - pour some water from the jug into the two bowls
   - scoop a small quantity of soap flakes into one bowl and swish the water to dissolve the soap
   - rub the dirty cloth in the soapy water
   - squeeze out as much water as you can from the soapy cloth
   - put the cloth into the bowl of clean water to rinse
   - wring it out again
   - peg the cloth on the clothes line or drape it over the clothes rack to dry.

3. Once you have been through the whole process, let a volunteer have a turn at washing their own item (the others can watch and advise).

4. Encourage the children to look out for small items in the setting that may need washing – napkins, small hand towels, face flannels, dusters, cloths for wiping tables, dolls' clothes.

Flag day

Make and fly some party flags.

What you need:

- Plastic carrier bags – a few white ones, but some coloured too
- Scissors
- Permanent markers
- Stout string
- A stapler
- Dowel or garden canes

Top tip ⭐

If you can link this activity with a special day it will give your party flags more meaning.

Taking it forward

- Show pictures of flags you think your children will recognise – national flags, the Red Cross, the Jolly Roger, soccer teams. Talk about what these flags mean, and why they are flown.

- Get children to design their own flags, including their favourite colours and things they like. Pin them up and get the children to talk about their own flag, or guess whose is which.

- Make a class or group flag with the children's names on it.

- Make or buy some national flags and fly them in your outdoor role-play area.

What's in it for the children?

This activity gives children experience of recognising and talking about symbols, and thinking about what they mean and how they are used.

What to do:

1. Tell the children that you are going to make some party flags. What shape will they be?

2. Help the children to cut flag shapes from the carrier bags (squares, rectangles, triangles).

3. Use the markers to draw patterns and other markings on them. You could fringe the edges of some of them.

4. Cut a long piece of string and staple the flag shapes securely to it.

5. Go outside, hang up your flags and watch them flutter in the wind.

6. Fix some of your flags to sticks and have a parade waving them.

➕ Health & Safety
Take care with scissors and sticks.

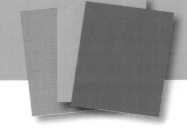

Wind machines

Make fans for a hot day.

What you need:

- All sorts, colours and types of paper cut into A4 sizes
- Felt pens, coloured pencils or crayons
- Sticky tape or a stapler

Top tip ⭐

Stick to fairly thick paper – thin papers (tissue and some wrapping papers) are not as effective.

Taking it forward

- Make alternative fans by sticking a paper plate to some stout dowel. Wave it about to create a breeze.

- Shake a sheet to create a much stronger wind.

- Set up a discovery table with bought fans (hand and electric kinds), and try some out. Pin up the homemade fans close by for a stunning display.

- Cut out fish shapes from thin tissue or newspaper, put them on the floor and move them along by wafting the fans. Make this into a race.

What's in it for the children?

This activity helps to develop hand and eye co-ordination and manipulative skills.

Health & Safety

As always, when things are being waved about, be aware of the threat to eyes.

What to do:

1. Talk with the children about what you can do to cool down on a hot day. Tell them you're going to make your own wind machines. Give out – or let them choose – a sheet of paper each.

2. Using the felt tip pens or coloured pencils decorate both sides of the paper, or stick on a collage of wrapping papers.

3. Show the children how to make a fan by beginning at the short side of the paper, and making a fold about 2 cms wide. Turn the paper over so that the folded part is underneath the rest of the sheet and fold back the edge again.

4. Repeat these two steps, folding and turning, until the entire sheet of paper is folded.

5. Hold one end of the folded paper firmly and fix at the end with sticky tape or a staple.

6. Holding the fixed end, open out the fan.

7. Flip your wrist rapidly to create a breeze. Move the fan near your face to feel the air waft.

Catch a falling leaf

Leaves are not all the same – study them and enjoy the wonder of autumn.

What you need:

- An area with a range of different trees
- Suitable outdoor clothing
- Bags for collecting leaves
- Magnifying glasses
- Camera or smartphone

Top tip ⭐

This activity should be done in autumn and is linked to 'Look what the wind's blown in!' on page 28.

What to do:

1. With the children, plan a simple walk which will take you where there are different types of trees. You could draw maps together of the way you plan to go and what the children think they will see.

2. Keep to small groups of children for each walk, and take plenty of adults to help with looking and talking.

3. Go to your chosen area and find the trees. Touch the tree trunks and feel the bark.

4. Gaze up into the branches of the trees and watch them move in the wind. Watch the leaves falling from the trees. See if you can catch them (some people say that if you do it brings good luck).

5. Each group should make a pile of leaves. Each child takes a leaf from their pile and runs round the area to find someone else who has a similar leaf. The match could be by kind, shape, colour or size.

6. When a match has been found these children can stand together until everyone has found a match. Now play again.

7. You could play the same game with seeds and nuts if you find these.

Taking it forward

- Pretend to be a tree – stand in a clear space, gently swaying in the wind. Stretch arms up high for branches. Someone walks around, pretending to be the wind and lightly touches each tree. When touched, the tree comes alive and bends and sways. Eventually everyone will be moving in a forest dance.

- Make a tree circle by standing around a tree trunk and holding hands. Move slowly round, maintaining the circle.

- Talk about why some trees lose their leaves. If appropriate for your group, introduce the different terms for those that do and those that don't.

What's in it for the children?

This is a good opportunity to use and explore the mathematical concepts of shape, colour and size.

Bike wheel spinner

Make a spectacular spinner that will give pleasure in any weather.

What you need:

- The front wheel from a bicycle, complete with axle but without the tyre: it must have offset spokes, as on adult bikes, and a larger wheel will work better

- A piece of wood about 5 cm x 2.5 cm and 10 cm longer than the radius (i.e. the distance from the hub to the rim) of the bike wheel

- Wide plastic tape, clear or coloured

- A drill and a drill bit, the diameter of the axle

- Screws, tape or cable ties to fix your windmill to a post or wall

Top tip ⭐

The key is finding a good wheel that spins easily. Try parents, or cadging from your local bike shop. You can see a video showing how to make the wheel on YouTube – http://www.youtube.com/watch?v=ITXqFe_aG1I

Taking it forward

- Use the wheel to help you judge the strength of the wind and to contribute to your weather recording.

- Tape the wheels of bikes and trikes, and watch the patterns as they're ridden around.

What's in it for the children?

Some aspects of this activity will be difficult, but overcoming the challenges will give a great sense of achievement.

What to do:

1. Tell the children you are going to make a spinning wheel, and introduce the wheel and the other equipment. An adult will have to do some of this, but the children should be able to do a lot of it, with help and supervision.

2. Lay the wheel on its side, and wrap the tape around the spokes. The aim is to fill the gap between a pair of adjacent spokes with tape, miss out the next pair, put tape between the next pairs, and so on until you've done alternate pairs of spokes all the way round the wheel. Use clear tape if you like, or mixes of coloured tape for a dramatic result.

3. When you've taped the spokes, put the wheel aside. Take the piece of wood and drill a hole the diameter of the axle, 4 or 5 cm from one end.

4. At the other end, drill two smaller holes for the fixing screws.

5. Now find a place outside to fix the wheel. Ideally it should be somewhere where you can see it from indoors. Screw the piece of wood horizontally to an upright post, the edge of a wooden shed, or perhaps the edge of a brick wall (although you'll need to drill the holes with a masonry drill and use plugs to screw into).

6. The wheel will go round in even the lightest winds, and if you've used coloured tape it will make a spectacular display.

✚ Health & Safety
Take the usual precautions with tools.

50 fantastic ideas for rain, wind and snow

Parachutes

Make the most of the wind with a parachute.

What you need:

- A parachute, or big square of lightweight fabric
- A large, open space outdoors
- A camera or smartphone

Top tip ⭐

A proper parachute works very well, but you can get just as good results with a big square of silk, rayon, nylon, polyester or other light fabric. Don't try this activity when it's too windy.

Taking it forward

- Take photos and make a collection of pictures of parachute activities. Use them to get children talking about what they saw, heard, thought and felt.
- Make some miniature parachutes for small world figures. Throw them into the air, or toss them out of an upstairs window, and see how they fly.

What's in it for the children?

Working together, listening carefully, co-ordination – this activity practices all these and more.

What to do:

1. This is good on a nice, breezy day in the summer. Get out your parachute(s) or large squares of fabric and ask the children to stand around the edges. Make sure everyone has a piece of the fabric to hold on to.

2. Start with some warm up activities:

   - Touch the sky! – everyone raises the parachute up as high as they can.
   - Touch the floor – bring it down to the ground.
   - Wave! – flap the parachute in the air.
   - Foot inside! – everyone puts one foot under the canopy.
   - Walk! – the group walks round in a circle.
   - Underneath! – all children run under the parachute.

3. Next, try these games:

   ### Weather report
   Pretend the parachute is the sea. Call out 'the sea is rough' or 'the sea is calm', or 'it's very windy', 'slight breeze', 'flat calm'. The children make the parachute behave according to the report given. Once they get the hang of it, get children to take turns being caller.

   ### Merry-go-round
   Children turn their bodies sideways and hold the parachute with one hand. They walk around in a circle, making a merry-go-round. For variety, children can hop, skip, jump, etc. Clap your hands to give them a beat to keep. Call out 'change' for them to go in the other direction.

   ### Tag
   Lift the parachute high into the air. Call out two children's names. The chosen children must trade empty spots by running under the parachute before it comes down on them.

   ### The wave
   Children put their hands up, one after another, in order – creating a Mexican wave effect.

   ### Flying saucer
   Everyone lifts the parachute in the air and takes one step forward. Call 'Let go!' and watch the parachute as it slowly floats to the ground.

Winter wonderland
Take a walk in the ice and snow to see how things are different.

What you need:

- Boots, coats, scarves, gloves, hats
- A few spares in a bag or backpack
- Some rewards to keep the walkers motivated, such as bananas, raisins, apples and corn chips
- A camera or smartphone

Top tip ⭐

Save this activity for a really cold spell, when there's ice and maybe even snow. Don't forget to wrap up warm.

Taking it forward

- Some animals get through the winter by going to sleep (introduce the word 'hibernate' if it doesn't come from the children). Make a list of animals that do this. Use the internet for help.
- How do creatures that don't hibernate cope? Talk about migration, and some of the birds that do it.
- Look in books and online for pictures of really cold places. Who lives there and how do they cope?

What's in it for the children?

This activity gives children the chance to think and talk about the features of wildlife in winter.

 Health & Safety
Wind-chill affects children more than adults. Be aware of this and don't keep the children out for too long.

What to do:

1. Tell the children you are going for a walk in the snow. What should they wear?

2. If you've already done the wet weather or the windy weather walks (or both), go to the same places. Talk with the children about what you saw then – things should look very different this time.

3. When you're out, look at the frost on trees, bushes, grass, rooftops and railings. Talk about what you see.

4. Find some ice-covered puddles. Test them to see how thick the ice is. What sound does it make as the ice breaks?

5. Is the wildlife asleep in the winter? Are some plants dead? Look for traces of animals and birds – prints, and holes in snow. Look at tree branches for traces of new buds coming for next season.

6. Take some photos to make a record and to talk about later.

7. When you get back, load your photos onto your IWB, if you have one, and talk about the walk. Get the children to describe what they saw and heard, and to talk about how the cold, ice and snow felt.

8. Make a winter display of their drawings and the photos. Label it with some winter words.

freeze it, melt it
What happens when things freeze?

What you need:

- Containers of various shapes and sizes
- Ice trays
- Balloons, rubber gloves
- Food colouring

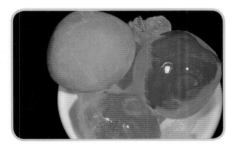

Top tip ⭐

Wait for a cold spell, and check the weather forecast for a very cold night.

Taking it forward

- Talk about how freezing enables us to keep food fresh.
- Make ice patterns by putting seeds and berries in jar lids, filling them with water and putting them outside to freeze (or in the freezer).
- Investigate icebergs, in books and online.

What's in it for the children?

This activity encourages investigation and observation, and raises questions such as 'What do you think?' and 'What will happen if?'

 ## Health & Safety

Make sure that the children's hands don't get too cold.

What to do:

1. Leave several containers of water in your outdoor area to be 'discovered' by the children the following morning. (If you can't make use of natural freezing, fill an ice cube tray and leave it in a freezer.)

2. Help the children to remove the ice from the tray or containers and explore it. What does it look like? What does it feel like? Can they pick it up? How long can they hold it?

3. Talk about what happens to the ice when it's been in the warm for some time. Where does it go? What does it turn into? Could it be made back into ice again? How?

4. Provide a selection of different containers (pots, tubs, trays, balloons, even rubber gloves). Fill them with water and put them outside overnight (if cold) or in the freezer.

5. The next day bring them in and enjoy the discussion which goes on as the children investigate them. Drop the ice balloon or ice gloves into a bowl of water. Do they float or sink?

6. Add food colouring or paint to the water before freezing it. Mix a selection of different coloured ice cubes in a bowl of water and see what happens as they melt.

7. Make a note of the words the children use and the comments they make to build a word bank of ice words.

Everyone is different

The shapes of snowflakes are endlessly fascinating.

What you need:

- Outdoor clothing
- Snow – the fluffier the better
- A large piece of black fabric
- A magnifying glass
- Squares of white paper
- Scissors
- Pipe cleaners
- Beads
- A camera or smartphone

Taking it forward

- Spread white glue on one of your snowflakes and sprinkle salt over it. Put it aside to dry – the salt will add a sparkle. Crumple some polystyrene into small pieces to make snow scenes for small world play.

- Talk about how the snow affects us and how we cope with it. There may be snow reports on TV news, or there are plenty of snow videos on YouTube.

- Make up some songs and rhymes about the snow.

What's in it for the children?

As well as studying one of nature's wonders, these activities offer practice in cutting out and following instructions.

Health & Safety

Take usual care with scissors, and don't keep the children out in the cold for too long.

What to do:

1. Go out in the snow. If you're lucky it will be fluffy snow with large flakes. Let some of the flakes fall onto the black material and look at them through the magnifying glass. Take some close-up photos if you can.

2. When you've spent some time looking at the flakes, take the children indoors and talk about the snowflakes you've seen. Use your photos, or get some pictures of snowflakes from the internet. Ask the children about the flakes – can they describe the shapes, colours and patterns? Encourage them to come up with words that fit the flakes.

3. Tell the children you are going to make some paper snowflakes. Demonstrate first, following the steps below. It's tricky, and some children will need help.

   - Fold the paper square in half to make a rectangle.

   - Fold the rectangle in half to make a smaller square.

   - Rotate the square so that it's positioned like a diamond. The folded tip, which is the paper's centre when unfolded, should be at the bottom.

   - Fold the diamond along the middle to make a triangle.

   - Keeping the paper folded, snip away shapes along the edges. It will look better if some of them are irregular.

   - Carefully unfold the paper to see the snowflake you have made.

   The children's first attempts might not be impressive, but with a bit of practice they should become more skilled. If you'd rather use ready-made templates for your snowflakes there are plenty online, or you can download some from **http://www.firstpalette.com/tool_box/ printables/snowflake.html**

4. Talk about your snowflakes. How are they like the real ones? How are they different? Make some snowflake garlands and hang them in your setting.

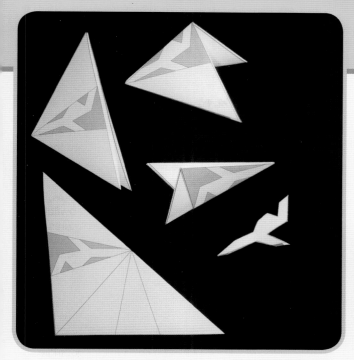

Top tip ⭐

Make some paper squares beforehand by folding the top right hand corner of an A4 sheet diagonally across to meet the paper's left edge, and cutting off the piece left over.

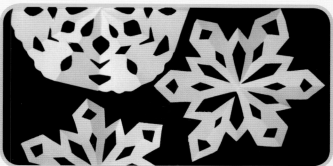

5. To make bead snowflakes:

• Twist two pipe cleaners together to make an X.

• Twist another pipe cleaner across the middle of the X to make a shape with six evenly spaced 'arms'.

• String different sizes of beads on one of the arms. Fold the end of the pipe cleaner over the last bead.

• String beads in the same way along the other five arms.

• Wind two pipe cleaners around the ends of the 'star', adding beads as you go.

• Tie a piece of thread to the end of one of the arms and hang your snowflake up.

Snow dance

Explore the snow and imitate it how it falls.

What you need:

- Outdoor clothing
- A camera or smartphone
- Space indoors to move about

Top tip ⭐

If you don't have the opportunity to watch snow falling outside, try to find some videos of snowstorms on the internet.

Taking it forward

- Make snowflakes out of white netting fabric: cut strips about 25 cm long and 10 cm wide and tie two strips together in a knot in the middle to make a simple prop that children can wave about while they dance.

- Some people think that there are dances that will make it snow. Enter 'snow dance' as an internet search to see some of them.

- Play 'Follow the leader' and run, hop, skip in the snow – the children have to follow what you do.

What's in it for the children?

This is a good movement activity which gives children lots of stimulus to use their imaginations.

➕ **Health & Safety**

Wind-chill affects children more than adults. Be aware of this and don't keep the children out for too long.

What to do:

1. Go outside in the snow. Watch how the flakes fall. Chase the snowflakes and try to catch them.

2. How do the snowflakes move? Spin and whirl, imitating the way they fall.

3. Back indoors, tell the children you are going to do a snow dance.

4. Start by getting the children to pretend they're walking through mud. They have to pull hard to get their boots out of the squishy mud. Can they pull their boots out of the mud?

5. Now the mud isn't so squishy, and there are lots of puddles. Stomp in the puddles and splash.

6. Now it's cold and all the puddles have turned to ice. Ice skate, forwards, backwards, on one foot. Turn and spin. Make patterns on the ice.

7. Now it's snowing. The snow is deep. Lift your legs high each time you take a step. Make patterns with your feet in the snow.

8. Imagine you have a sledge. Sit down and race down the hill on your sledge. Fall out of it and roll through the snow.

9. Roll a snowball to make a snowman. We will stack them one on top of the other. Now imagine now you're a snow person. How does it feel to be a snowboy or snowgirl?

10. The sun is coming out. What will happen to us when the sun shines on us? Melt very, very slowly until you end up as a puddle on the floor.

50 fantastic ideas for rain, wind and snow

Winter picnic

We're used to picnics in the summer, but a picnic in the snow is special!

What you need:

- **Cold weather clothing** (coats, hats, gloves, scarves, boots – try to have some spares)
- **Rugs or plastic sheets**
- **Spare bags** (for rubbish)
- **Paper or plastic plates and cups**
- **Handwipes**
- **Food and drinks** (warm if possible)
- **A camera or smartphone**

Taking it forward

- Try a more complicated picnic, taking a camping stove and doing some cooking.
- If you want to go down the Ray Mears route, build a campfire. Keep it up together in a circle of stones. Let the flames die down and cook things such as sausages or marshmallows on the glowing embers, on skewers. A bit of charcoal added will keep it going longer. Make sure any meat is thoroughly cooked before the children eat it.
- Find some picnic stories and songs (e.g. *The Picnic* by John Burningham, *The Lighthouse Keeper's Picnic* by Ronda Armitage and *Having a Picnic* by Sarah Garland).

What's in it for the children?

This activity gives lots of practice in planning an event.

 ### Health & Safety

Make sure that the children's hands don't get too cold.

What to do:

1. Plan ahead and choose your day. Gently falling snow is fun, but if it's any heavier it probably isn't – particularly if it becomes freezing rain or hail! Don't go too far away from your setting, so you can get back quickly if you need to.

2. Involve the children in preparing the food. Get suggestions from them, but the obvious choices will be crisps, sandwiches, small cakes, biscuits, bananas, apples. If you can manage a few warm things (sausages, soup, baked potatoes, pizza) they will be appreciated.

3. Take photos, and when you get back use them to help the children talk through the experience.

Top tip ⭐

Don't make it too complicated the first time. Snow reflects the sun, so if it's a sunny day use sunscreen.

Snow colours

White snow is lovely, but why not improve on nature with a little colour?

What you need:

- **Warm clothing**
- **Spray bottles with water and a few drops of green, red, yellow or blue food colouring** (you need five or six drops of food colour per cup of water)
- **Paintbrushes** (foam brushes work best for this)

Top tip ⭐

Dry snow is best, although you can get some good results on packed snow and ice too.

Taking it forward

- If you've made a snowperson you can colour it, giving it a red nose, pink cheeks, and coloured buttons.

- Make an ice hanging. Pour water into a cake tin and stand a cup in it (off centre – weigh it down with a stone). Add a few drops of different food colourings. Don't stir the water up or move it – just leave it there to freeze. Once frozen, carefully tip out the ice and hang it up outside by threading a ribbon through the hole where the cup was.

- Add some food colouring to pots of water and leave them outside overnight to freeze.

What's in it for the children?

We're not used to coloured snow. This activity gives children the chance to see things in a new way.

 Health & Safety

Children's fingers will get cold if their gloves get very wet.

What to do:

1. Find an area of clean snow and let the children have the spray bottles, taking a colour each. Allow some time for experimenting.

2. Suggest to the children they work together, combining their colours to make patterns.

3. Give them some challenges. Can anyone write their name in the snow? Can anyone make a picture of a flower? A car? An animal?

4. Paint the snow with brushes.

5. Try painting snow that lies along the tops of branches, on fences, on window sills. Before you go in, have a snowball fight!

50 fantastic ideas for rain, wind and snow

Snow patrol
Look out for tracks and trails in the snow.

What you need:

- A clipboard or notebook and pen
- Bird food or breadcrumbs
- A guide to birds and animals (if possible)
- Binoculars (if possible)
- Camera or smartphone

Top tip ★

If you can, visit the place you're planning for the children to go beforehand, and scatter some bird food or breadcrumbs to attract wildlife.

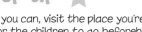

Taking it forward

- Use wheeled toys to make tracks in the snow.
- Make a bird table: tie thin ropes to the corners of a wooden vegetable box, hang it from a tree, add some bird food and watch them visit.
- Look on the internet for images of tracks in the snow. Can you tell what animal or vehicle made them?
- Make some tracks or prints in clay or play dough.

What's in it for the children?

This activity gives practice in observing and recording.

 Health & Safety

Take the usual precautions when you take a group of children out.

What to do:

1. Tell the children you are going out to look for tracks and trails in the snow. Remind them of tracking in the mud, if you've done that activity.

2. Tell them that they'll have to walk very carefully so they don't frighten the creatures or wipe out the tracks.

3. Visit an open space – in your own grounds, a park, or a path – where there are likely to be animal or bird prints.

4. Talk about what you find, and record it with the camera and by making drawings.

5. Find some tracks of people, perhaps near your setting. Can you tell who walked there? Was it a child? A woman? Or a man?

6. Put on your biggest pair of boots or snow shoes and walk through the snow. The children have to walk in your footsteps.

7. An adult cuts some shapes out of thin foam and ties them to the bottoms of their boots. The adult walks through the snow to a hiding place. The others have to follow the tracks to find him or her.

Snow people

Make mini snowmen, snow animals and snow angels.

What you need:

- Plastic trays
- Small world figures and toys

Top tip ⭐

A snowman doesn't have to be huge. In some way small ones are more fun.

Taking it forward

■ Make snow slime and play with it! Mix two cups of white glue in a bowl with one and a half cups of very warm water. In another bowl, combine a small teaspoon of borax with a cup of very warm water. Mix both bowls well and then combine the contents. Mix with your hands for a few minutes while the snow slime forms. Adding glitter will make it really special.

■ Frozen white rice is an alternative fake snow.

■ Make a small world habitat for a polar bear.

What's in it for the children?

This activity gives children the chance to experience snow play in unusual ways. The snow slime and the frozen rice are very tactile to explore.

➕ Health & Safety

Children's fingers will get cold if their gloves get very wet.

What to do:

1. Make a snowman. It doesn't have to be a big one. You can make mini snow people out of snowballs. Make lots of them, and some snow animals.

2. Children love making snow angels. They lie on their backs in the snow and wave their arms up and down. Use food colouring to make faces on the angels and to give them yellow wings.

3. Take some plastic trays outside. Make a snow scene for your small world people. Include ski slopes, snow houses, ponds for skating, cars, tractors and diggers. You can use an ice cream scoop to make even smaller snowmen.

4. Your outside snow scene will melt, so make one indoors using fake snow: mix eight cups of flour with one cup of baby oil and work it well together till it's like soft sand. Put it in the freezer to give it the authentic chilly feeling.

Snow Olympics

Organise your own Olympic games in the snow.

What you need:

- **Hula-hoops**
- **Plastic cups**
- **Plastic mats or trays** (for sledging)
- **Clipboard, pen and paper** (for keeping score)
- **Camera or smartphone**

Top tip ⭐

You might want to spread the games out over several sessions - if the snow lasts long enough.

Taking it forward

- Play 'freeze-tag'. The person who is 'it' must try to tag the other players. Once a player is tagged, they must freeze in place. Once all the players have been tagged, someone else takes over as 'it'.

What's in it for the children?

These games keep children active and make them think.

 Health & Safety

Remember that children sometimes don't realise when they're getting cold.

What to do:

1. Get the children ready to go out for some games in the snow and ice. Some of these are competitive and some collaborative:

   - Make targets in the snow with hula-hoops or by drawing in the snow with a big stick. Have a snowball tossing contest to see who can score the most points.

   - Hide a ball or a carrot or something brightly-coloured in the snow. Give the children clues to find it.

   - Play snow mini-golf: bury plastic cups and see if you can get a hole in one. If you don't have a golf club or if hitting the ball is too hard, try rolling the ball in.

   - Find a good icy patch and see who can slide the furthest on a mat or tray.

   - Play noughts and crosses: draw the 'board' in the snow with a stick and use twigs and stones for the x's and o's.

   - Pair up for a contest to make the biggest snowball. It ends when the teams can no longer roll their entries or when you run out of snow (or get too cold). You could set a time limit.

   - Play 'Snittles': stand some sticks in the snow and try to knock them down by throwing snowballs at them (three goes each).

2. If you want the games to be competitive, list them and keep score.

3. Take lots of photographs to talk about later.

The Gruffalo's Child
Play a story in the snow

What you need:

- *The Gruffalo's Child* by Julia Donaldson
- Card mask shapes for each child: either a mixture of creatures or all Gruffalo's child
- Felt pens
- Scissors
- Elastic
- Warm clothes and boots

What to do:

1. Download some mask templates of the Gruffalo's child from Google Images. It would be easier to have the Grufffalo's child masks for all the children, but you could add the Gruffalo and the Snake, Owl, Fox and Mouse masks to make the story more complicated. Often, being in the snow seems to affect children's ability to concentrate!

2. Explain that you are going to go out in the snow for a snowy story, but first the children need to make their masks.

3. Show the children the mask shapes and help them to make them, by cutting out and colouring the templates and fixing some thin elastic to fit their heads.

4. Now get dressed in warm coats and boots. Don't forget to bring the book, and put on your masks.

5. Let the children run about a bit first, particularly if it is the first snow of the winter.

6. Gather them together somewhere that is dark, like the cave — maybe a doorway, or under the climbing frame. Start to tell the story. You could be the Gruffalo, and the children could all be the Gruffalo's child. If you have more characters, they need to spread out through the 'deep dark woods' and wait till you bring the other 'Gruffalo's child' children near.

7. Tell the story as you move round your play area, looking for footsteps in the snow, and stopping when a Gruffalo's child finds another animal.

8. As the story gets nearer to the end, make sure you all return to the cave so the Gruffalo's child feels safe.

9. Go inside and warm up. Maybe you could make some hot chocolate!

✚ **Health & Safety**
Make sure you are all warmly dressed. Children get cold very quickly.

50 fantastic ideas for rain, wind and snow

Taking it forward

- Sing some nursery songs, changing the words for the snow; you could try something like this:

From 'Teddy Bear, Teddy Bear'	'Snow, Snow Faster'	'The North Wind Doth Blow'
Gruffalo, Gruffalo when it snows,	Oh where oh where has the Gruffalo gone,	The North wind doth blow
Gruffalo, Gruffalo, touch your toes,	Oh where oh where can she be?	And we shall have snow,
Gruffalo, Gruffalo, touch the ground,	With her hair all brown and her teeth sticking down,	And what will the Gruffalos do then,
Gruffalo, Gruffalo turn around.	Oh where oh where can she be?	Poor things?
Gruffalo, Gruffalo, all alone,		They'll sleep in the cave,
Gruffalo, Gruffalo, run back home,		And keep themselves warm,
Gruffalo, Gruffalo, run into the cave,		And cuddle up tight till it's spring,
Gruffalo, Gruffalo, you were very brave!		Good thing!

What's in it for the children?

Telling a snowy story in the snow really brings it to life.

Top tip ⭐

Telling a story before you go gives a purpose to the walk, and keeps the children moving as you act out the story together.

Frozen world

Even things children know well look different when they're frozen.

What you need:

- **Small containers of different sizes**
- **Tea lights** (battery operated ones preferred)
- **Jar lids**
- **Buttons, beads, sequins**

Top tip ⭐

When you're freezing things in containers, start by spraying the surfaces lightly with cooking oil.

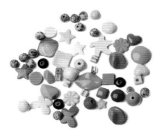

Taking it forward

- Draw some pictures of how you made the ice lanterns. Arrange them in the right sequence. Make a display with them.

- Make some bird feeders in jar lids. Instead of using buttons, beads, etc., add bird seed to the water. When they're frozen put them on a bird table or flat surface. Watch and photograph what happens.

What's in it for the children?

This activity gives practice in observing, making and sequencing.

 Health & Safety

Children's fingers will get cold if their gloves get very wet.

What to do:

1. Go outdoors and look at frost on leaves and branches. What does it look like: sugar? salt? icing? Encourage the children to come up with describing words, and use them in sentences.

2. Make ice lanterns. Get two containers of different sizes. Put a little snow in the bottom of the larger one and pack the smaller one into it. Pour cold water into the space between the two containers. Add a few drops of food colouring and place outside to freeze (or leave in the freezer). When it's frozen, remove the inner and outer containers. Put tea lights in the laterns and use them to decorate the outside area (use battery operated electric tea lights – they're safer for children and they won't blow out). These are great at Christmas.

3. Take some shallow containers (jar lids are good) and fill them with water. Add buttons, beads, sequins, berries and seeds and put them outside to freeze. What do they look like when they are frozen? What happens when the ice melts?

Snow toffee

Make some snow toffee on a cold day.

What you need:

- A cup of brown sugar, 55 g butter, 1 teaspoon of vanilla
- A tray for the snow
- Ice cubes (optional)
- A heavy saucepan
- A cooker
- Paper bags (to put the toffee in)

Top tip ★

This is a great activity for a really cold day.

Taking it forward

- Carry out a poll to find out which sweets are the most popular. Make a chart to show who likes what.
- Sweets are good in moderation, but too many are bad for the teeth. This is an excellent time to do some work on teeth and the importance of cleaning them.

What's in it for the children?

Following a recipe is good practice in understanding instructions and doing things in a methodical order.

✚ Health & Safety

If you collect snow outdoors, be careful where it comes from and make sure it is freshly fallen. Keep children well away from the pan and the hot, sticky liquid.

What to do:

1. Talk about sweets. Ask the children which ones they like. Do they know how they're made? Do they like toffee? Tell them they are going to make some special toffee in the snow.

2. Collect together the ingredients and double them if you have a lot of children, or if they want to take some toffee home.

3. Go outside to collect fresh snow on a tray. Make sure that the snow you collect is very clean! Put it in the freezer (or, if it's cold enough, cover it and leave it outside in a shady spot). If you can't find clean snow (or don't fancy using what you find outdoors!) make some. Put a few ice cubes in a blender, one at a time, and whisk them till they're the texture and consistency of snow. Have the snow ready before you go any further. Keep it frozen until stage 6.

4. Mix all the ingredients together in a bowl and give it a good stir. Scrape the mixture into a heavy pan.

5. Adults only: put the pan on the cooker and bring the mixture to a rolling boil. Boil till it's at the 'hard ball' stage (i.e. when dripped into very cold water it turns into a firm ball).

6. While this is going on and when the mixture is almost ready, the children can get out the snow they've collected or made. When the mixture has reached the desired temperature, drizzle it over the snow. The mixture will cool and harden very quickly. The toffee can be eaten within a few minutes, or stored in a cool, dry place for a few days.

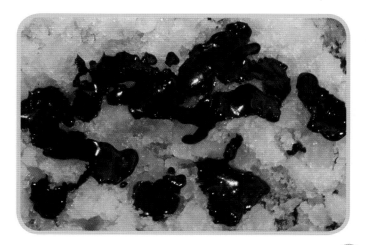

A snow den

What you need:

- Empty plastic containers *(ice cream containers are fine)*
- Trowels, shovels, spades
- Cooking oil spray
- Large plastic tray
- Plastic tunnel

Top tip ⭐

It's not often you get enough snow to make a snow den, so grab the chance when you can.

Taking it forward

- Allow the children to take some plastic furniture into the snow den.

- If you have enough snow – and children – make several snow dens. Connect them together with plastic tunnels.

- Use books and the internet to find out about people who live in igloos.

What's in it for the children?

This provides good practice in working together.

 Health & Safety

This can take quite a long time, so start early, keep the children warm, and have some refreshments handy to keep them going.

What to do:

1. Clear a space, putting the snow you move in a pile.

2. Lightly spray the insides of the containers with cooking oil. Pack the snow into the containers and squash it down. Slightly wet snow works better than fluffy snow. If the snow is fluffy, you can add a little water as you pack it.

3. Tip out the snow bricks and arrange them in a circle, leaving a space for the door.

4. Make another layer of snow bricks on top of the first one. Stagger the joins, like a brick wall.

5. Continue adding layers. If you want the genuine igloo shape, step each layer of bricks in very slightly as you go.

6. Stop when you get to a suitable height. Place the plastic tray on top and heap more snow on it (NB. Don't try to make a snow roof without the tray — you don't want children buried in snow if it collapses.)

7. Make a proper entrance — you can use a plastic tunnel and pack snow round it.

8. Enjoy playing in your snow den. It should last well and will survive mild thaws.

Homemade snowflakes
Grow a snowflake in a jar!

What you need:

- String
- A wide–mouthed pint jar
- White pipe cleaners and wire cutters or pliers
- **Blue food colouring** (optional)
- **Boiling water** (with adult help)
- **Borax** (if you can't find it locally, look online)
- A pencil

Top tip ⭐

If you do this activity in the morning the children can check on the snowflake in its solution during the day to see it forming.

Taking it forward

- Try making some coloured snowflakes by mixing different food colourings into the borax solution.

- Make some more pipe cleaner snowflakes, lay them flat, roll paint over them and print with them.

What's in it for the children?

This activity gives an opening to talk about what crystals are and how they form.

 Health & Safety

Borax is safe in normal use, but take care no one inhales the powder, and avoid contact with skin or eyes.

What to do:

With a little kitchen science you can create long lasting snowflakes that are as sparkly as the real ones.

1. Cut a white pipe cleaner into three equal pieces. Twist the pieces together in the middles so that you have a six-pronged star shape.

2. If your points are not even, trim the pipe-cleaner sections so that they're all the same length.

3. Attach string along the outer edges to form a snowflake pattern. Tie one end of a short piece of string to the top of one of the pipe cleaner arms, and tie the other end to a pencil (this is to hang the snowflake from).

4. Fill a wide–mouthed jar with boiling water (job for adults only!). Mix borax into the water, one tablespoon at a time, and stir until dissolved, (don't worry if there is powder settling on the bottom of the jar). Use three tablespoons of borax per cup of water.

5. It looks better if you add a very little blue food coloring to give the snowflake a bluish hue.

6. Submerge your pipe cleaner snowflake in the jar with the pencil resting on the lip and the snowflake freely suspended in the solution.

7. Wait overnight, and by morning the snowflake will be covered with shiny crystals. Hang it to dry, then hang it in a window as a sun-catcher or use as a winter time decoration.

Brilliant bubbles

Have some fun on a freezing cold day when there is no wind in the air.

What you need:

- 1–3 tablespoons glycerine
- ¼ – ⅓ cup washing-up liquid (biological washing-up liquid doesn't work)
- 1 gallon water

On 'bubble day' you'll need:

- Wool or soft string
- Bendy straws
- Pipe cleaners, coat hangers or soft wire
- Baking sheet
- Plastic wrap
- Clean sock
- Containers, such as big plastic boxes or a large bowl

Top tip ⭐

If you make the bubble mixture a day or so before, it will make better bubbles.

Taking it forward

- Experiment by bending the wire into different shapes – e.g. a square, a heart, a diamond.
- Add food colouring to bubble mixture and let the bubbles land on paper. Watch what happens.

What's in it for the children?

Blowing bubbles is fun, and blowing them to make ice balls on a really cold day is a quick, satisfying way of seeing what the cold can do.

✚ Health & Safety

Make sure children are very warmly dressed, particularly their hands, feet and heads. Don't stay out too long.

What to do:

1. Create some bubble wands before you make the mixture. You can use wire coat hangers bent into shape and taped to make a handle, pipe cleaners twisted together, or loops of soft wire (electric wire, craft wire or garden wire; it should be sturdy enough to hold a circular shape, yet not so strong that it is difficult to bend).

2. Make some bubble mixture and leave it to stand overnight. This improves the bubbles. If the children are helping, you may want to make double quantities, so you can use some immediately.

3. In general, non-windy days are best, and the things you can do with bubble mixture are endless.

4. To make ice bubbles when it is very cold, blow a bubble outside and then catch it on your bubble wand. Wait a few moments while it freezes- it will turn into a lovely crystal ball before it shatters!

5. Make a bowl full of bubbles by blowing through straws into a washing-up bowl of bubble mixture.

6. Make a bubble bounce. Put a clean sock like a glove over your hand, blow a bubble and catch it on the glove, then bounce the bubble into the air.

7. Make bubble honeycomb. Wrap a rimmed baking tray in cling film, pulled taut. Lift a corner of the cling film and pour 1/3 of a cup of bubble mixture onto the baking tray, then blow through a straw to make bubbles until the space between the cling film and the baking tray bottom is filled with flattened bubbles. You'll find that the bubbles form perfect hexagons.

50 fantastic ideas for rain, wind and snow

Treasure hunt

Children will love the challenge of being set to find things.

What you need:

- Cards
- Collecting bags

Top tip ⭐

You don't have to use a different 'treasure' each time. Shuffle the cards and repeat them.

What to do:

1. This is an outdoor activity that you can do in any weather. Beforehand, write on cards the things you want the children to collect. What they are will depend on the weather and the season, but you might include a shiny pebble, a big leaf, a fir cone, a flower, a beetle (collected carefully), moss, grass, a twig, an acorn, a conker, a berry, a seed, a piece of ice, an icicle, etc.

2. Take the children to a contained area (e.g. in your own grounds, or the corner of a park).

3. Start by reading the object on the card and setting the children the challenge of finding it and bringing it back to you. They should be working collaboratively in groups.

4. When they've done this a few times, give them more than one object to find at a time.

5. To warm the children up, turn it into a race for individuals ('Who will be the first person to bring me a …?').

6. Make the challenge harder by introducing numbers – 5 leaves, 7 pebbles, 12 twigs, etc.

Taking it forward

- Hide some objects in the area chosen for your treasure hunt (a coloured ball, a crayon, a badge) and include those.

- Get the children to hide a treasure and draw a map showing where it is.

What's in it for the children?

This activity gives children practice in observation, concentration, searching and working together.

⊕ Health & Safety

Some berries and plants are poisonous. Make sure the children know to keep all plants – and the fingers that have touched them – away from their mouths.

Weather station

Tune into the weather when planning this activity.

What you need:

- Fabric streamer
- Broom handle
- Straight sticks or dowel
- Plastic water bottle
- Plastic sticky tape, glue
- Small squares of plastic (make your own, cut from yoghurt pots)
- Waterproof markers
- A compass
- Clipboards, notebooks and pens for recording

Top tip ⭐

Check the equipment regularly yourself. It will be very disappointing for the children if it doesn't do what they expect.

What to do:

1. Talk with the children about how hard the wind blows or how much rain falls, and ask them to suggest how you can find out. Tell them you are going to make a weather station together.

2. Measuring the wind:

- The easiest way to measure the wind is to tie a fabric streamer near the top of an upright post.
- It will help to know the wind direction. Take two sticks and fix them together with tape or glue to make a cross.
- Mark the points of the compass – N, S, E, W – on four of the plastic squares and stick them to the ends of the cross.
- Using tape or glue, fix the cross on top of the broom handle. Work on the sawn–off end, not the rounded one.

3. Measuring the rain:

- Talk about what you could use to collect the rain. Help them decide which will be the best to use – a clear plastic cup or the trimmed bottom of a plastic drinks bottle will work well. This will be your rain gauge.
- Mark lines a centimeter apart on the gauge.

4. Setting up:

- Go outside with the children and decide where to set up your station. Position the wind streamer in the open, if possible where you can see it from your windows. It will be easier if you can fix it to a fence or climbing frame upright. You can use cable ties to do this. Use the compass to line it up.
- Place the rain gauge in the open, clear of any overhanging trees or bushes. You could sink it in a bucket of sand to stop it blowing away.
- Help the children to design a simple recording chart for recording the daily wind and rainfall. They could think up symbols to show different weather conditions.
- Ask questions that make the children think. Are all days windy? Does it always rain on a windy day?

5. Don't forget to check and record the weather (and if necessary empty out the rain gauge) every day.

50 fantastic ideas for rain, wind and snow

Taking it forward

- Children who are able may like to use a computer to check the daily weather forecast for your area, or to look at the forecast in the local paper.

- Make a weather box. Get a large box or container and put your weather resources in it so they're always available.

What's in it for the children?

This activity gives children practice in making equipment, and recording data.

✚ **Health & Safety**

Take usual care with scissors, and don't keep the children out in the cold for too long.

All dressed up

When it's *too* wet, windy or cold to go out, try this indoor activity.

What you need:

- **Old magazines and brochures** (travel brochures are especially good)
- **Scissors**
- **Gluesticks or glue**
- **Backing paper or a scrapbook**

Top tip ⭐

Start collecting magazines and brochures well ahead, so you have plenty for the children to use.

Taking it forward

- Get dressed in your outdoor gear and sing one of the songs you have made up.
- Draw and paint a rainy, windy or snowy background for some more cut out pictures.

What's in it for the children?

Selecting, cutting out, sticking, and sorting are all useful activities for children.

Health & Safety

Remind children to use and carry scissors safely.

What to do:

1. Talk with the children about what to wear in the rain, wind and snow.

2. Go through the magazines, cutting out and collecting pictures of people dressed for those conditions.

3. Use the pictures to practice sorting. Sort for wind, rain, snow or all three. Sort according to the colours people are wearing, or whether they are children or adults. See if the children can suggest any other ways you could sort them.

4. Make a scrap book or collage of clothing for bad weather.

5. Make up some songs about clothes and dressing, to tunes the children know ('The Wheels on the Bus', 'Here We Go Round the Mulberry Bush', etc.). There are suggestions in *The Little Book of Clothes and Fabrics* (Bloomsbury).

The weather band

Make your own instruments for a rainy, windy orchestra.

What you need:

- Tins and plastic containers of various sizes
- Tape (duct and plastic), elastic bands
- Rice, popcorn (unpopped), small beads
- Cardboard tubes
- Foil
- Plastic sheet and balloons

Top tip ⭐

Make a collection of tin, tubes, etc. in advance so you have plenty for the children to pick from.

Taking it forward

- Practice on your instruments and put on a performance for another group.
- Wait for a wet or windy day, go to where you can get a good view of the outside and play your instruments along with the weather.

What's in it for the children?

Keeping time and rhythm is an important skill, and it's great to practice this on your own instruments.

What to do:

1. If you made the windchimes, remind the children of what you did then. Tell them you are going to make a weather band, using instruments that make weather sounds.

2. Make shakers by putting dry rice in tins and plastic containers.

3. Make drums by stretching plastic across the tops of large tins and fixing them with elastic bands or tape. Instead of using plastic, cut the top off a balloon and stretch it across. You can make drums from old paint cans.

4. Tape a funnel to the end of a length of plastic pipe and blow into it to make a wind noise.

5. A large sheet of hardboard or tin can make a thunder machine.

6. Small beads swished about in a biscuit tin make a sound like the sea.

7. Make a rain stick from a mailing tube (or tubes from rolls of wrapping paper work well). Duct tape over one end. Take a 60 cm length of foil, crush in longwise to make a long cylinder and wrap it in a spiral around a broom handle to make a coil. Take the foil off the broom handle, then push it into the tube and secure it at both ends with tape. Drop two tablespoons of dry rice and one of unpopped popcorn into the tube and tape up the end.

8. Decorate your instruments with coloured paper, paint and markers. Add some streamers.

9. Put on some music and play along on your weather instruments. Go out and play in the wind, snow or rain.

In the right gear

Help children to get ready for going outside more quickly and easily.

What you need:

- Items of clothing and shoes with buttons, zips, Velcro, laces, buckles and any other types of fastening you can find.

Top tip ⭐

Practise doing up and undoing a button and a zip yourself before introducing this activity to children. Concentrate on the skill needed and you will see how complex these actions are!

Taking it forward

- Make some 'fastening cards' so that the children can practise fastening buttons, zips and laces. Put a doll's cardigan or small zipped jacket onto a body shape cut from stiff card. Cut a shoe shape from stiff card, punch in some holes and thread the holes with a lace.

- Make a collage person out of stiff card and add lots of different fastenings stuck on with strong glue.

What's in it for the children?

Fastenings are important features of the children's clothes, and they need to learn how to use them independently.

➕ Health & Safety

Remind the children not to put items in or near their mouths: small parts can be a potential choking hazard.

What to do:

1. Introduce an item of clothing, such as a cardigan with buttons. Talk about how to do up and undo the buttons when wearing a cardigan. Ask the children to find the buttons on their own clothes and on others'.

2. Have the children try on different items of clothing with buttons. Help them to do up and undo the buttons if they need it.

3. Talk about other types of fastening, e.g. zips, hooks, buckles and Velcro. Show the children how they work and listen to the sounds they make. Try on the different items of clothing with these fastenings.

4. Find children who have coats with zips, hooks or toggles, or shoes with Velcro or laces, and ask them to show the other children. Can the children think of other items that have these fastenings (bags, purses, pencil cases, belts, sandals, jeans)?

5. See if the children can find or suggest other ways of fasting things – e.g. press studs, hooks and eyes, safety pins. Find some examples and practise with them.

# Leaf art

Printing fascinates children and offers quick and colourful results.

What you need:

- A bag to put the leaves in
- Paint
- Brushes
- Sheets of painting paper

Top tip ★

This activity is best done in the summer, when the leaves are big enough to use and are fresh and pliable.

What to do:

1. Tell the children that you're going out to collect some leaves. You want as many different types, sizes and shapes as they can find. Collect them in the bag, making sure they're not squashed.

2. Back indoors, cover your work area with a mat or newspapers, and get out the paint.

3. Tip out the leaves and pick one for the first print.

4. Position a leaf with its under-side facing up and paint on its entire surface. This is best done by placing the leaf on a paper towel and painting it with a fairly large brush. Make sure the entire leaf is covered but don't use too much paint – you don't want it to run over the edges and get on the unpainted side of the leaf.

5. Carefully turn the leaf over so the painted side is on the paper. Gently press the leaf onto the paper (not too hard, you don't want to squash it).

6. Slowly peel back the leaf to reveal a beautiful leaf print.

7. Repeat the process, using different colours and other leaf shapes.

8. Talk about your work, and make a display.

Taking it forward

- Try painting different colours on one leaf to create a rainbow leaf print.
- Try printing on paper plates and other surfaces, as well as on paper.

What's in it for the children?

This activity gives children practice in thinking through processes and creating patterns.

✚ **Health & Safety**

Keep an eye on what the children are picking up.

Snail city

Snails and other minibeasts are easy to find in the summer in the rain.

What you need:

- A large plastic tray (a builder's tray is ideal)
- A plastic or fabric cover for the tray
- Collecting boxes
- Snails
- Damp soil or compost
- Water
- Logs, leaves, twigs, soil, rocks
- Cabbage and lettuce leaves
- Magnifying glasses

Top tip ⭐

Snails like cool, moist, dark places. Leave out citrus rinds or cabbage leaves under upturned plant pots to attract snails so they're easy for the children to find.

Taking it forward

- Gently put some snails on black paper and watch them make silver trails as they move.

- Get a piece of glass or perspex, and cover the edges with carpet or duct tape to make them safe. Gently lift some snails onto the glass and then look at the other side of the glass. You will be able to see how they move and how their mouths work.

What's in it for the children?

This activity will help develop children's awareness of the natural world, and gives practice in looking after creatures.

➕ Health & Safety

Tell the children to wash their hands after they've been handling creatures or soil.

What to do:

1. Tell the children that you're going to make a habitat for snails in a tray. Ask them to tell you what they think the snails would like to have in there.

2. Tell them you're all going out to collect snails. Get them suitably clothed and equip them with collecting boxes. A damp day is best. Look under stones, behind things and among dead leaves.

3. While you're outdoors, collect stones, twigs, sticks, logs, etc. for your snail tray.

4. Put some damp soil in the tray and let the children arrange the logs, stones, etc. Add some leaves for the snails to hide in, and a couple of lettuce leaves for food. Snails need water, so make sure they have plenty.

5. Offer some magnifying glasses to watch the snails, and a camera or smartphone to record what they do. Use a magnifying glass to see how they eat, how they move, how their eyes work. Watch what happens when they retract and extend their eyes, and as they go into their shells.

6. Find a sheet of plastic or fabric to cover the snail house at night — otherwise the snails may escape and eat things!

7. At the end of the activity, take the children and return the snails to their natural habitat.

Pack it in
Get the right clothing for the right weather.

What you need:

- A selection of **hot weather clothes** (t-shirts, shorts, cotton skirts, sandals, sun hats)
- A selection of **cold weather clothes** (mittens, winter coats, scarves, woolly hats, fur lined boots)
- A box or basket for the clothes
- Two large suitcases
- Large luggage labels, felt pens and pencils

Top tip ⭐

Start off by exploring the clothes. Name them, talk about them and let the children dress up in any items that are easy to put on.

Taking it forward

- Introduce some holiday–related items such as wash-bags, postcards, travel pillows, old train and plane tickets, maps of cities, cameras.
- Use books, travel brochures and the internet to make a list of some hot places and some cold ones. Where would Great Britain fit?

What's in it for the children?

This activity involves thinking about the characteristics of items and planning.

What to do:

1. Explain that you are going on a hot, sunny holiday and a cold, snowy holiday and you need to pack one case for each holiday.

2. Invite a child to pick an item of clothing from the basket. Discuss with the group whether it should go in the 'hot holiday' suitcase or the 'cold holiday' suitcase.

3. Talk about each item of clothing, linking the clothes with the children's own memories of hot and cold weather. Help them to focus on the features that make it suitable for a hot or cold weather holiday - short sleeves or long sleeves - thick woolly material or thin cotton - fur lined or thin fabric?

4. Show the children how to fold each item of clothing and pack them neatly in the suitcase. Once your packing is finished, ask the children to help you close the suitcases.

5. Give the children some large luggage labels to write and draw on. Tie them to the suitcases and add these and some blank ones to your role-play area.

Let's have lunch

Rain, wind and snow all have a part to play in growing the food we eat.

What you need:

- A plastic tub or trough and a plastic tray
- Kitchen roll
- Scissors
- Bread and butter for the sandwiches
- Seeds

Top tip ★

Check the growing plants regularly. Make sure they have enough water, and look out for slugs and snails, which can destroy your crop overnight.

Taking it forward

- Grow beansprouts by putting a few mung beans in a plastic jar, covering the top with muslin and securing with an elastic band. Water the seeds daily by filling the jar with water and tipping it out again through the muslin. The sprouts should be ready to eat in five or six days.

- It's obvious why plants need rain, but see if you can find out why they need the wind and the snow.

What's in it for the children?

This activity gives children experience of growing and tending things, and contains some good sequencing concepts.

✚ Health & Safety

Tell the children to wash their hands after they've been handling creatures or soil.

What to do:

1. Discuss with children how the rain helps things to grow. Tell them you are going to grow some food to eat.

2. Start by planting some lettuce. It's best to grow them from seed (sprinkle the seeds in compost in a large trough or tub, or use a grow bag, and water well) — but if this is a problem buy small living lettuce from the supermarket (sometimes sold as 'cut and come again'). Leave them somewhere they'll get plenty of water and light. Check them every day. The seedlings should begin to sprout within seven to fourteen days.

3. When your lettuce is getting to a good size (three or four weeks after sowing), start some mustard and cress by sprinkling the seeds onto damp kitchen roll in a tray. Sow the mustard first and put the tray in a light place indoors (best to avoid direct sun through a window – it will dry out too quickly). Three or four days later, sow the cress. Keep the tray moist.

4. When the lettuces have reached a reasonable size, harvest them. By this time the mustard and cress should be 5 or 6 cm tall. Harvest with scissors.

5. Make sandwiches using the produce you have harvested. Have a picnic to eat them.